# IN THE NATIONAL INTEREST

General Sir John Monash once exhorted a graduating class to 'equip yourself for life, not solely for your own benefit but for the benefit of the whole community'. At the university established in his name, we repeat this statement to our own graduating classes, to acknowledge how important it is that common or public good flows from education.

Universities spread and build on the knowledge they acquire through scholarship in many ways, well beyond the transmission of this learning through education. It is a necessary part of a university's role to debate its findings, not only with other researchers and scholars, but also with the broader community in which it resides.

Publishing for the benefit of society is an important part of a university's commitment to free intellectual inquiry. A university provides civil space for such inquiry by its scholars, as well as for investigations by public intellectuals and expert practitioners.

This series, In the National Interest, embodies Monash University's mission to extend knowledge and encourage informed debate about matters of great significance to Australia's future.

Professor Sharon Pickering
President and Vice-Chancellor,
Monash University

# JENNIFER RAYNER

# CLIMATE CLANGERS: THE BAD IDEAS BLOCKING REAL ACTION

*Climate Clangers: The Bad Ideas Blocking Real Action*
© Copyright 2024 Jennifer Rayner
All rights reserved. Apart from any uses permitted by Australia's *Copyright Act 1968*, no part of this book may be reproduced by any process without prior written permission from the copyright owners. Inquiries should be directed to the publisher.

Monash University Publishing
Matheson Library Annexe
40 Exhibition Walk
Monash University
Clayton, Victoria 3800, Australia
https://publishing.monash.edu

Monash University Publishing brings to the world publications which advance the best traditions of humane and enlightened thought.

ISBN: 9781922979636 (paperback)
ISBN: 9781922979650 (ebook)

Series: In the National Interest
Editor: Greg Bain
Project manager & copyeditor: Paul Smitz
Designer: Peter Long
Typesetter: Cannon Typesetting
Proofreader: Gillian Armitage
Printed in Australia by Ligare Book Printers

A catalogue record for this book is available from the National Library of Australia.

The paper this book is printed on is in accordance with the standards of the Forest Stewardship Council®. The FSC® promotes environmentally responsible, socially beneficial and economically viable management of the world's forests.

For Luka, Lola and Jude,
who will live with the consequences
of the choices we make now.

# CLIMATE CLANGERS: THE BAD IDEAS BLOCKING REAL ACTION

This is not one of those books that mounts an urgent and important case for doing more to tackle harmful climate change. There's no question we must, but you already know that.

You know that with the world having warmed by around 1.2°C so far since the mid-eighteenth century, we're all suffering for it. Maybe you're a parent who is watching extreme heat and dangerous bushfires turn the carefree summers you remember into a season of anxiety for your kids. Or you're a community leader grappling with the wild swings between flood and drought that are costing homes, lives and livelihoods. You might be a business owner paying more for insurance than you can afford and worrying about whether the next big storm will be the one that shuts you down for good.

Whoever you are, you've lived through years of escalating risks and wilder extremes, so you know that climate change is harming us here and now. You can see that worse is on the way if we don't deeply cut harmful carbon pollution this decade by rapidly phasing out coal, oil and gas. And because you know this, because *so many of us* know this, it's perplexing and infuriating that this isn't yet happening at anything like the speed and scale we need.

Governments around the world are making promises and plans—few outright climate deniers are holding the reins of power right now. But so far, all those pledges still have us on a pathway that would see the global average temperature rise by almost 3°C in the decades to come.[1] This is a simply ruinous level of heating—for us and for the millions of other species we share this planet with. At half that temperature rise, we already risk reaching tipping points—think melting ice sheets and dying coral reefs—that could accelerate dangerous heating past the point of no return.

Even as we are building more clean solar and wind power than ever before, we're still digging up, shipping out and burning fossil fuels in record amounts. We've started buying electric vehicles, but carbon pollution from our cars keeps edging up anyway. We're letting whole industries off the hook with the great catch-all excuse that their emissions are 'hard to abate'. Frankly, the steps we're taking to build a clean energy economy and move beyond fossil fuels are far too incremental for the existential nature of this threat. Why?

Three bad ideas are holding us back from taking real climate action commensurate with what this challenge demands. They are perspectives or conceptual frameworks that usually go unsaid, because they're so deeply embedded in how politicians, policymakers, business leaders—maybe even you—see and understand the world. They're like an in-built set of blinkers that obscures the full view of what is necessary and possible for tackling harmful climate change right now. Let's put these three bad ideas under a spotlight, so we can cut through their falsehoods and break their persuasive spell.

The first is that our economy must keep growing as fast as it always has while we decarbonise. There is an almost fetishistic belief that to be a thriving and prosperous country, economic growth—measured in terms of gross domestic product (GDP)—must march on quarter over quarter, year after year. This idea is a major handbrake on the root-and-branch renewal we need to decarbonise Australia's economy. GDP growth is simply the wrong way to track the health of a clean economy. Obsessing over it distracts us from what we really need to do to build one—and build it quickly.

The second is that 'net zero' accounting can keep global heating within survivable limits. This is the notion that some companies and industries, like our fossil fuel giants, can keep polluting as usual if someone, somewhere else, does things that 'offset' their pollution on paper. The whole concept of offsetting is based on a flawed understanding of how carbon pollution cycles through our natural world and causes dangerous heating. But big polluters have grabbed it with both hands and shoved it centrestage in the public conversation about what counts as climate action. This is muddying the waters when it comes to understanding what it will really take to halt global warming, and why we must hold big emitters accountable when seeking genuine and permanent cuts to carbon pollution.

Finally, there is the idea that taking the necessary action on climate change now will cost us more than we can afford. This assumes that the price of action will be greater than the cost of inaction. With global climate-fuelled extreme weather disasters now costing over US$143 billion *a year*,[2] it's obvious that this is wishful thinking. It's worth taking a look at just how much unchecked climate change could cost us, together with the trillions we'll have to spend in the decades ahead to prop up fossil fuels if we don't quickly phase them out. We need better ways of tracking the costs on both sides of the ledger, which will help us distinguish between dumb investments and smart ones.

In my day job, I spend more time than is possibly healthy talking to politicians, policymakers and other Very Serious People about climate action. I sit around over-large tables in airless parliamentary offices, where the clocks loudly tick away like reminders of political mortality, while people tell me it's simply not possible to move at the speed and scale the science demands. They say we can't end fossil fuel exports while there's still someone out there who'll buy them. They say we need 'least cost abatement' to get business and markets moving. They say of course they'd *like* to do more, but it'll have to wait until better budgetary times. Whether they realise it or not, a lot of very smart people with their hands on the levers of power are bound up by these three bad ideas. They shape how decision-makers understand the problem of climate change, and how they imagine its solutions. Just as consequentially, these ideas lead them to overlook or rule out the actions we really need to get on with.

The longer we leave these ideas unchallenged as the world keeps warming around us, the more dangerous they become. We have to dismantle these ways of thinking and build better ones in their place, so that we can get on track with the real action that matters now.

## GROWING PAINS

What does the wellbeing of Australians have to do with how much a Californian oil CEO earns, or the annual share dividend paid out to a French bank? If you listen to politicians and economists, the answer is quite a lot. Economic growth—most often measured in terms of GDP—is their collective shorthand for the health of our economy.

Standard economic thinking says that when a country's economy gets bigger, the living standards of its citizens improve. So if GDP is growing briskly month on month and year after year, all is well with the world. If GDP growth is sluggish—or worse, going backwards—this is taken to mean storm clouds are gathering on the economic horizon, and hard times lie ahead for households and businesses. Governments are on tenterhooks in the hours before each quarterly GDP update is released, waiting for word that growth is still ploughing on and their record on economic management remains intact.

But if you dig into what GDP actually measures, the focus on it as *the* main yardstick for the health of our economy seems a little bizarre. That's because the standard GDP metric that gets touted by the media and obsessed over by markets doesn't directly relate to our living standards at all. It's simply a measure of production, adding up the value of everything that was produced and sold for money in Australia's market economy during a given period of time. The value of everything *produced and sold for money*—not the value of what we all earned or even got to use within our borders. GDP calculations make no distinction between economic benefits seen here at home and those which are enjoyed by foreign companies funnelling their profits overseas. That's where Chevron CEO Mike Wirth and the financial services giant BNP Paribas come in.

Australia has long been a resources-heavy economy. The mining and energy sector contributes about 10 per cent of national GDP through digging up and shipping out fossil fuels like coal, oil and gas, as well as different types of minerals.[3] We are the world's third-largest fossil fuel exporter, with carbon pollution from these exports being more than double Australia's domestic emissions.[4] A lot of the corporations that extract and ship these products are foreign multinationals, like the oil and gas giant Chevron. In 2021, it reported over $9 billion income from its Australian business, against a direct economic contribution of just under $2 billion.[5] That direct contribution comes through things like employing workers, sourcing materials and equipment from other local businesses, and paying state resources royalties. But the multibillion-dollar gap between this and its revenue? Much of that is profit which mostly goes offshore, back to Chevron's parent company in the United States. This—together with paying almost no Australian company tax—helped it post a US$15.6 billion profit for its operations globally the same year.[6]

Chevron's global arm parcels some of that out to Mike Wirth and other top company executives for their yearly salaries and bonuses. It splits some of it with major shareholder BNP Paribas and international investors as a dividend on their stocks. Chevron might even invest some of this profit in new fossil fuel projects in other countries, so that it can start the whole extractive cycle all over again. These are also the end beneficiaries of the production that gets measured when we tally up Australia's GDP—not just our own community. When Treasury types proudly tout strong economic growth as a sign of our prosperity, they're conflating what's good for us with what's good for firms and investors around the globe.

This phenomenon isn't unique to Australia. But it's a bigger issue for us than it is for many other countries because of the extent to which our GDP is boosted by the production and sale of raw commodities, by huge firms that are mostly based overseas. Selling coal, oil and gas has proved to be one of the most profitable activities in human history. Fossil fuel corporations are still posting record results even as the need to rapidly wean the world off their polluting products becomes increasingly urgent. For example, the oil colossus Shell announced a record US$39.9 billion profit for 2022, off the back of skyrocketing global prices caused by Russia's invasion of Ukraine and the energy supply crunch that followed. That was a particularly good year for the company—one person's war profiteering is another's juicy share dividend, after all. But even in the ho-hum year immediately before that, Shell still posted a US$19 billion profit.[7]

When the fossil fuel sector does well, this flows directly into our GDP figures because the value of its production is calculated on the price its products fetch. So while there have been ups and downs in global prices over the years, it still means that for decades now, Australia's GDP has been juiced by the contributions of coal, oil and gas. We've been counting these as a core part of our own economic story, even though a big chunk of the financial benefits has gone offshore. This helps explain why the idea of phasing out Australia's fossil fuel exports causes politicians and economists to break out in hives: it would make our economy look smaller, when bigger is always assumed to be better.

Our exports aren't the only way in which fossil fuels contribute to this particular measure of economic growth. Household consumption—the things you and I buy in our daily lives—makes up roughly half of Australia's GDP. Because household consumption is such a large share of what gets counted in this measure, the way we meet our daily needs matters a lot for how big our economy is calculated to be and whether or not it's growing.

Take home energy as an example. The average annual electricity bill ranges from around $1200 to over $2000 per household around Australia. At the moment, most Australians buy their power through a retailer, which purchases it from generators in the wholesale electricity market. At the end of 2023, around 60 per cent of Australia's electricity was still sourced from coal or gas-fired power stations, so those generators also need to buy those raw materials to produce the electrons they sell down the line. Each of these transactions—household to retailer, retailer to generator, generator to coal or gas company—creates economic activity which shows up in our GDP figures. The same goes for buying petrol for our cars, powering our businesses with gas, and anywhere else we engage in market transactions involving fossil fuels.

Spotted the snag yet? As soon as a family paying $2000 a year on their power bills installs solar panels and a battery, something important changes. They've gone from sourcing electricity through a long chain of regularly recurring market transactions, to producing it themselves. The purchase, installation and maintenance of their solar kit will be counted in our standard GDP statistic, but a big chunk of their ongoing energy costs has permanently dropped off the books. That family isn't worse off; their standard of living has actually improved because they're spending far less on power bills and have more control over their energy needs. There will also continue to be plenty of jobs in the energy sector, because installing and maintaining all that personal power infrastructure requires plenty of skilled workers. This is a clear example of how economic activity that currently adds up to GDP can disappear, while households and workers keep thriving.

To deeply and permanently cut harmful carbon pollution, we need to take steps like this everywhere, weeding coal, oil and gas out of every corner of our economy. That means transforming everything from how we source energy and move around, to how we make things and use land. I'm not talking about a slight tweak to the systems we've relied on since the industrial revolution—this needs to be a wholesale renovation. As we do this, our economy will almost certainly grow more slowly than it has in the past. Measured GDP may even go backwards at times. This isn't something we should actively seek out, like proponents of the sackcloth-and-ashes degrowth agenda propose for rich countries globally. It's just something that's going to happen naturally, because we'll be spending less on things that cost us a lot today and selling less fossil fuels for the benefit of overseas firms.

In our current frame for thinking about the economy, flat or falling GDP growth is treated like a four-alarm fire—a crisis. Pursuing policies and actions that may throw off its ascent is considered to be nothing less than political suicide. But with profound decarbonisation, our economy can contract on this metric while Australians keep prospering, and, in fact, become better off in important ways.

Truly. When we start looking at the kinds of steps we need to take to permanently cut carbon pollution, we can spot plenty of examples like the home energy one. Switching from driving a petrol-powered car to walking, cycling or using electrified shared transport, will shift market-based transactions on expensive fuel to free or lower-cost ways to get around. Household spending on transport will fall, but Australians will still be able to get where they need to go. Profits may no longer be flowing to foreign car brands or our old friends at Shell, but more workers will be needed in transport services here at home.

The increased adoption of plant-based diets; switching from owning every tool and toy to accessing them when we need them through the sharing economy; recycling more materials and products at home, in our built environment and in industry—the list goes on. For all these examples, the GDP measure may show a decline when Aussie living standards are actually unchanged or even improving. Deeply and permanently cutting emissions doesn't just mean switching from one polluting technology to a directly equivalent zero-emissions one. It means reshaping the economic structures that underpin how we make and consume, because those structures are a big part of what got us into this climate mess in the first place.

The obsession with GDP growth is a product of that old paradigm. Continuing to chase it in the new one simply makes no sense.

Worse, assuming a zero-emissions economy must keep growing, like our fossil-fuelled one has done, stacks the deck against going all-in on decarbonisation, because it's simply not going to deliver that. This is a fact we particularly need to face in relation to Australia's exports. To tackle harmful climate change, we have to stop extracting and exporting the coal, oil and gas that are fuelling it. Demand for these products will decline as countries around the world take their own steps to cut emissions. The 2023 *Intergenerational Report* noted that if the world successfully achieves the 2015 Paris Agreement goal of limiting global warming to 1.5°C, the demand for Australia's thermal coal will fall to just 1 per cent of what it is today. Even in a scenario where countries collectively hold warming to 2°C—which brings with it plenty of other environmental and economic risks—coal demand will still halve.[8] The same is true for Australian gas exports as new energy and technology solutions replace this fossil fuel in homes, businesses and industry. Likewise crude oil exports, which will increasingly be rendered obsolete by the rapidly rising uptake of electric vehicles.

In order to have a thriving economy that delivers for our community, do we need to replace every dirty dollar that comes from fossil fuel exports today? If we're locked into the GDP mindset, we do. That's why politicians and policymakers are currently casting around for other silverware to sell.

The popular story right now is that we can keep growing by selling other countries the materials they need to decarbonise. The Albanese government's *Critical Minerals Strategy* notes that the world will need forty times more lithium and twenty times more cobalt and graphite—among other key minerals Australia is rich in—to deliver on emissions-reduction commitments. Treasurer Jim Chalmers has described critical minerals as Australia's 'opportunity of the century'.[9] That's certainly true, but there's no guarantee that critical and rare earth minerals will turn out to be as insanely and enduringly profitable as fossil fuels have been. The thing about coal, oil and gas is that as long as you're relying on them to heat homes, power industry or move millions of vehicles around, you have to keep buying them. Enormous quantities of them. Significant amounts of critical minerals and rare earths will go into producing the solar panels, wind turbines, heat pumps and batteries that will be the backbone of our zero-emissions energy and industrial systems. But it's not necessarily the case that demand for these goods will be as high and insatiable as it has been over decades for our fossil fuels, or that they'll always fetch the premium prices they do today. Once they've been built, solar panels and wind turbines last quite a long time. A lot of these materials are also reusable and recyclable in a way that fossil fuels aren't.

That's OK, because the relationship between the GDP value of something and its contribution to Australian living standards isn't one to one. A fossil fuel company and a critical minerals miner may employ the same number of people at comparable wages. They may also source similar levels of goods and services from other local businesses, and pay equal rates of company tax and royalties (we wish!). If these companies sell their respective products overseas for very different income, their contribution to GDP will differ, *but their real economic contribution to our community is exactly the same*. It is these direct benefits for Australians that we need a decarbonised economy to deliver, not a substitute for every single dollar fossil fuel exports currently add to the statistical size of our economy.

How is all this holding us back from doing what's necessary for our climate? While we remain fixated on

GDP growth, we can't imagine turning off the rivers of black and blue gold until something else shows up that can fully take their place. But that sets the bar too high for what clean alternatives can, and need to, deliver. It's simply not likely our GDP *will* keep growing at historic rates if we fully transition coal, oil and gas out of our domestic economy and exports. So if our politicians, business leaders and media keep treating this as the implicit test, we'll remain stuck where we are today: unable to see a pathway that satisfies it, and as a result, putting off a full phase-out of fossil fuels. That's why we must front up to the fact that GDP is past its use-by date, and start tracking the health of our economy differently.

Defenders of 'green growth' vigorously dispute this. Progressive politicians and economic institutions prefer the idea of decoupling. Historically, any growth in economic activity has tended to come with an increase in harmful carbon pollution. That's because the things we do and produce to make money depend on burning fossil fuels. Decoupling suggests that as we shift to renewable energy and new, clean industries, GDP can keep rising while emissions fall. Advocates for this idea point to the fact that dozens of countries around the world have achieved some degree of decoupling in recent years. For example, between

2013 and 2019, the United Kingdom managed to grow its economy by an average of 2 per cent a year while cutting emissions by over 3 per cent a year.[10] This is the pollyanna picture of decarbonisation: a world where emissions go down and economic growth keeps ticking ever upwards.

The problem is that the rates of decoupling achieved so far won't cut carbon pollution anywhere near as much as is needed in the time now left to avoid climate catastrophe. One study which looked at decoupling in rich countries estimated it would take on average another 223 years for them to reach net zero on current trends.[11] This is time we simply don't have when the world is working towards net zero by 2050, and a strong body of science says we need to hit that milestone much earlier. Other research highlights that much of the decoupling seen in rich countries has been achieved by outsourcing high-polluting activity to other, poorer countries.[12] That's no way to cut emissions globally, because every country ultimately needs to get to net zero. Decoupling is a seductive idea, but it's not clear we can do it to the extent needed—in all countries—to prevent more harmful climate change.

We need better indicators than GDP to tell us if our economy is delivering for Australians, ones that aren't weighed down by the outsized legacy of fossil fuels and the profit expectations of foreign multinationals. The good news is that they're already out there.

### Employment and Investment: the Tests that Matter

The number-one metric we should care about when it comes to the health of our economy is employment. Having a decent job when you need one is the single biggest factor behind individual financial wellbeing. Creating enough work for everyone who wants it, across our community, is the foundation stone of shared prosperity. With a decent job, we can meet our daily needs, support those who rely on us, and feel secure about our future. In a thriving economy, everyone gets to share the benefits of good work, with one person's spending creating another's salary in a virtuous cycle of productive activity.

It's often assumed that GDP growth drives employment, but the evidence for this is more mixed than you might think. Research by the International Monetary Fund and other organisations has shown that the amount of jobs growth associated with an increase in GDP varies widely across the world—ranging from significant upticks in employment to no or negative changes.[13] For rich countries like

Australia, the research generally finds a positive relationship between GDP growth and employment, but not a particularly strong one.

Thinking back to the example of Australia's fossil fuel exports during 2022, this makes sense. Our GDP jumped because multinational coal and gas companies were able to sell their dirty wares for eye-watering prices during a global supply crunch. But they didn't need extra workers to reap these profits, so this spike didn't deliver any new jobs. We'd see the same thing if, for example, a manufacturer fully automated their production line and sacked all their flesh-and-blood workers: GDP would keep rising but workers wouldn't get the benefits.

So the first and most important test we should set for a clean economy is whether it's delivering good jobs for everyone who wants one. There's every reason to believe we can comfortably pass this test as we take the urgent steps needed to slash harmful carbon pollution in the coming decade.

The Australian Bureau of Statistics' labour market survey shows that in late 2023, around 82 000 Australians were directly employed in mining, producing and manufacturing coal, oil and gas products.[14] That's about 0.6 per cent of all employment, highlighting just how small the contribution of these industries actually is on this measure. They have to be phased out entirely, because mining more fossil fuels is incompatible with a safe climate future. Another 87 500 people worked in our energy system, supplying electricity and gas to homes and businesses. Getting off gas will permanently end some of these jobs, but many will simply change as workers move from coal-fired power generators to installing and running the infrastructure for making clean energy from wind and solar. Around 130 000 people sold cars and fuel, and a further 318 000 worked in road transport—moving goods around in highly polluting heavy trucks. Electrifying transport will update the kinds of vehicles some of these Australians work with, but to seriously cut carbon pollution, we need a big shift away from private and commercial road transport to shared, efficient modes like rail and other types of public transport. So all up, that means around 618 000 direct jobs are on the front line of fossil fuels throughout our economy—about 4 per cent of total employment today.

The most comprehensive analysis of Australian pathways to decarbonisation done so far, the Net Zero Australia study, estimates we'll need a skilled workforce in the order of 700 000 to 800 000 people to deliver net zero. This is significantly more than the number employed in those fossil fuel industries.[15] The Australian Government's Jobs and Skills Australia agency says we'll need up to 42 000 more electricians in just the next seven years to reach the national target of 82 per cent renewables.[16] We're already seeing this job creation in action, with employment in the electricity sector trending up in recent years as the rollout of renewables gathers pace. In Queensland's industrial centres, like Townsville and Gladstone, Andrew 'Twiggy' Forrest, Korea Zinc and other business titans are now racing to build massive wind and solar projects, powering a cleaner grid together with the production of renewable hydrogen and green metals. The same thing is happening from Gippsland and the Hunter to Collie out west: renewable energy and clean industry are delivering new jobs to start replacing those linked to fossil fuels.

We'll need a focused national effort to match workers in shrinking industries with good new jobs in growing ones, including updating skills and licences where this is needed. We'll also want to be smart about where new renewable energy and clean industry projects go, to share the benefits and make sure communities and regions right across the country keep thriving. But if we're creating enough good work for everyone who wants it, *this* is the most important sign of a healthy, clean economy. GDP can go up, down or sideways while we do so; the jobs are what counts.

This goal contains an important qualifier: enough good work *for everyone who wants it*. Australians have been feeling the pinch for years now, from colossal mortgage costs to rising food and grocery prices, unprecedented petrol bills and energy price shocks. We're struggling to cover it all, which can mean working much more than we'd like to. Taking on a second job, putting off retirement or returning to work before we're ready to be parted from our little ones are the hard choices more and more people are making. The data bears this out: the share of Australians working has been trending up pretty steadily (the COVID blip notwithstanding), and in late 2023 it was at the highest level on record.[17] That's mostly been driven by more people working part-time. If we had the same employment to population ratio today as we did in the 2010s, there'd be over 1.2 million *fewer* people at work.[18] Almost one million Australians now also have more than one job. The share of people with multiple jobs has been trending up progressively since things reopened after COVID, hitting its highest level ever in 2023.[19] This shift has tracked along with rising house prices and other cost-of-living pressures, suggesting more households need a second income, or an ongoing one later in life, just to make their budgets add up.

I've explained how cleaning up our economy is going to shift market-based transactions into free or lower-cost ways for people to meet their needs—for example, through renewable energy cutting power bills and shared transport ending price shocks at the fuel bowser. Economists often assume that when costs go down in one part of the household budget, spending just increases in another. Decarbonisation points to a more appealing potential. If Australians can meet their needs for less, this gives us the chance to work less if we choose to do so. Not everybody will, of course—some households will simply choose to spend more on other things they value. But gaining more time to spend with our kids while they're small, to volunteer in our communities, or to enjoy the retirement we've worked so long for, seems like a benefit plenty of people would be up for. Having the choice to work less because we need less to live comfortably is a bright spot we might miss if we think about jobs through the same frantic growth mindset we're currently applying to the whole economy.

Now, a decarbonised economy isn't a kibbutz, so someone has to create these jobs. That's why there's a second important metric we should use to track the health of our emerging clean economy: business investment. We need this to keep ticking along because starting new businesses and investing to expand existing ones leads directly to jobs. Research in Australia and internationally has consistently shown that new firms account for the vast majority of job creation, with small and medium-sized businesses being particularly strong engines of employment. Capital investment in things like an expanded production line or opening a new office also creates jobs because companies then need more workers to operate them. Contrary to the fretting over how new equipment will kill jobs by replacing everyone with robots, recent studies have found that capital investment in sectors like manufacturing has a positive overall impact on employment.[20]

I can hear your confusion: doesn't business investment just lead to growth, which then keeps feeding the beast of our GDP? It can do, but it depends a lot on the nature of the investment. A wind farm and a coal-fired power generator will both cost a big chunk of change to build. Once up and running, the renewables project has some production costs for things like its workforce, but its main input for making electricity is effectively free—you don't need to buy wind. That's a big contrast with the power plant, whose operators must pay for coal anytime they want to make a single joule of energy. Even if the electricity they produced then sold for the same price, the wind farm's total contribution to GDP would be less because it has fewer costs on the input side. (Of course, it *wouldn't* sell for the same price because this very lack of input costs means renewable energy is significantly cheaper than its fossil counterparts—another cut to GDP).

This is a particularly relevant example for decarbonisation, but the point holds true for any two businesses. A dollar of investment can produce different amounts of measured economic activity depending on the input costs for each business, the nature of the supply chain they sit within, and the price their products will fetch.

Since business investment matters because it's a driver of employment, we should care about investment that maximises job creation, not economic growth. One will sometimes create the other, but we can treat that as a happy by-product rather than the main game. As with GDP, business investment can create significantly different numbers of jobs depending on the sector it's poured into and the type of activity it funds. Creating new businesses in our trades and services sectors is especially jobs-rich, because in health care, education, and trades like plumbing and electrical, what a business sells *is* effectively people's labour. Investing in research and development can also deliver big jobs gains by unlocking the creation of new products and services, which then leads to the development of whole new industries or the expansion of existing ones. By contrast, fossil fuel mining stands out as one of the uses of business capital that creates the fewest ongoing jobs. A few years ago, the gas giant Woodside proudly announced it had poured $29 billion of investment into its fossil fuel projects in Western Australia; it made capital expenditures worth $2.3 billion in 2022 alone.[21] Having spent all these billions, Woodside employs a little over 3300 people in Australia—just 0.02 per cent of all jobs in this country.[22]

Put simply, we need a level of business investment that creates enough good jobs for everyone who wants one. That means the 'right' amount of investment will shift at times depending on who wants to work, and how much. We may not need to sustain a permanent forward march of growth on this metric any more than we do on GDP to meet this test. As we proceed with decarbonising our economy, we should focus on what business investment is delivering in direct outcomes for Australian workers, and avoid the old trap of thinking that bigger is always inherently better.

Embracing employment and business investment as the metrics that matter for a clean economy frees us up to think differently about what we need to do now. The questions that should guide our plans are: how do we create enough good jobs for everyone who wants one, and an environment that encourages businesses to invest in low- and zero-emission industries to do so? That's a meaningfully different matter than trying to keep our economy growing at historic rates forever. When governments and policymakers redefine their task in these terms, it unlocks new possibilities for phasing out fossil fuels much faster than we are today. Perhaps most importantly, it marks out an accelerated pathway to the end of coal, oil and gas exports through prioritising the creation of good alternative jobs for workers in these industries, not a one-to-one replacement of every dollar of GDP.

Jim Chalmers and the Albanese government are well placed to drive this shift. Since taking office in 2022, they have made employment a core focus of their economic story, hewing closely to the historic priorities of the Australian Labor Party. Climate Change and Energy Minister Chris Bowen never misses an opportunity to remind us that 'the world's climate emergency is Australia's jobs opportunity', or to spruik the new investment being poured into clean energy and industry. The government can build on this by clearly framing jobs and investment as the outcomes that matter as we work to cut carbon pollution, and directly acknowledging that GDP growth is no longer the right or relevant test for the economy we're building now. Ripping off this intellectual and political band-aid will encourage a new national story about success in a clean economy, and what can get us there.

## PAPER CUTS

Setting the right goals for our clean economy is the first step in building it. We then need to get our policies and plans lined up behind those goals so they're all pulling in the same direction. Right now, however, too much of the climate agenda—not just in Australia, but globally—is built around a flawed understanding of what it will actually take to halt and then reverse harmful warming. To accelerate the real cuts to pollution we need right now, we must to lay to rest the ill-founded idea that counting carbon offsets equals climate action.

If you tip out a binful of rubbish in your local park while your neighbour builds a backyard compost plant for theirs, is there still mouldy veg blowing around the swings? How about if you cheat on your partner while your best friend promises to be faithful to theirs—do you still deserve the cold shoulder at home? One more: if you skol six pints while the bartender makes some lemonade, does that mean you're sober?

The idea of 'net' accounting for social ills is completely ridiculous in almost all contexts, as these examples make plain. We don't want *any* rubbish piled up where our kids play, so the fact that someone responsibly dealt with theirs doesn't cancel out what's been dumped. The infidelities that matter in your relationship are (presumably) the ones that *you* commit. And however much soft drink your friendly barman makes, you're still going to have a headache the morning after downing so much beer.

At the moment, the prevailing way we think about, talk about and plan for tackling carbon pollution is based on logic as dodgy as that for being net sober. We've built an entire global accounting system for tracking progress towards net-zero emissions by 2050—collectively and in each of the 196 countries that have signed up to the Paris Agreement, Australia among them. This United Nations (UN)–led framework says we should aim to cut carbon pollution first and foremost, and then use offsetting actions to balance out the remainder. But it does not include any binding requirement for emissions to actually fall across the board. So big companies and some governments have piled onto the bandwagon of net zero achieved through carbon offsets instead, as a free ride to keep on polluting as usual.

Planning for net-zero emissions using this type of carbon accounting leads to some frankly insane outcomes. Like Woodside, Santos and Chevron claiming hand on heart that they're working towards net zero even as they're developing new fossil fuel projects that would see them pumping out pollution for decades to come. Or the mammoth global carmaker Toyota saying it can achieve this goal while continuing to sell petrol cars well beyond 2050. CEOs who make these kinds of claims should be laughed off their warmly lit AGM stages. Instead, investors applaud and laud their climate leadership.

The paradigm of net zero puts the focus firmly on requiring businesses to balance their carbon ledger. Australia's national task is defined in the same terms. But the only way to prevent more dangerous climate change is for carbon pollution to actually go down—not on paper in the annual emissions accounts, but in the real world, and for good. The idea that net-zero accounting can keep global heating within survivable limits is a furphy big corporations wield as a smokescreen for pollution as usual. We have to shake ourselves free of it and focus instead on the real action that will make a difference.

Over half those who took part in a 2023 poll conducted by The Australia Institute confidently asserted they understood how the idea of net zero and carbon offsetting works. But less than one in five people then correctly answered the follow-up questions to prove they really did.[23] So humour me in stepping through how all this is *supposed* to work, before we zoom in on why it doesn't.

The concept comes from the idea of sources and sinks. In our environment, some things are natural sources of emissions—like decaying plants and animals, volcanic eruptions, bushfires and burping cows. Others, like our oceans and forests, can absorb and store this carbon as a sink. Carbon flows between these sources and sinks in a continuous cycle, supporting a stable climate as long as they are relatively balanced. But starting with the industrial revolution, and especially since the 1950s, we've unleashed massive new sources of carbon pollution through burning coal, oil and gas. Because this hasn't been matched by new carbon sinks, there's been an increasing build-up of that pollution in the atmosphere.

And as we all know, this is smothering the planet like a heavy winter blanket and causing global heating.

Today's net-zero agenda has the kernel of a good idea at its heart: getting back to a balance between sources and sinks, so we are no longer producing more carbon pollution than can be safely absorbed by our natural environment. So far, so scientifically sound. The problem is that current plans to get there are heavily reliant on big polluters buying lots of carbon offsets, *not* on them stopping the flood of emissions into our atmosphere.

Carbon offsets are like tokens for carbon pollution. One token usually represents 1 tonne of $CO_2$-equivalent gas that has been reduced, avoided or removed from our atmosphere. There are myriad ways to create these tokens, some more trustworthy than others. The most common methods include tree planting and land restoration, because plants absorb and store carbon through photosynthesis as they grow; and renewable energy projects, because these avoid the pollution that comes with fossil fuel power. Currently, there's also a lot of hype about more high-tech carbon-removal technologies for scrubbing carbon from the atmosphere, with these being touted as potential future offsets if (and it's a big if) the technology can be scaled up and made affordable.

Projects that reduce or avoid pollution can register the creation of carbon offsets, and these are then bought by the corporations that create it. This lets those polluters—whether they're a coalminer, a steelmaker or a major retail brand—act like their emissions never happened. Sometimes they are required to purchase offsets by law, such as through the European emissions trading scheme or Australia's federal Safeguard Mechanism policy. Other times, companies buy them voluntarily to meet their own sustainability commitments and show what sterling corporate citizens they are. It's a massive business—in 2022, regulatory and voluntary carbon markets combined were globally valued at around US$100 billion, while covering less than a quarter of global emissions.[24]

Big corporations have been extremely successful at convincing governments and the community that buying these offsets effectively deals with their problematic pollution. In that same poll by The Australia Institute, 60 per cent of participants said they believed that when businesses offset their emissions, the overall amount of pollution in the atmosphere either stays the same or is reduced. Very few people correctly identified that absolute pollution levels are actually *higher*. To borrow from *The Usual Suspects*: the greatest trick fossil fuels ever pulled was convincing the world their pollution didn't exist.

There are three problems with the whole concept of offsetting. Remember these as the three Ss: storage, surplus and sources.

Storage is the most fundamental issue. Burning fossil fuels releases carbon that has been locked away underground for millions of years. Land-based offsets like planting trees or restoring wetlands do not permanently store that carbon away underground again. It cycles between land-based ecosystems, the ocean and our atmosphere, where it keeps building up over time. Think of it like dumping piles of new sand into an already full sandbox. Shaping some of it into castles and adding water to mix more of it into clay doesn't take the sand back out of the box again.

The carbon that does get absorbed by newly planted trees and land restoration is also very vulnerable to being released again because of fires, heatwaves, disease and droughts. That clay in our sandbox can quickly dry out and turn back into plain old sand. Fossil fuel pollution stays in our atmosphere for thousands of years, but the projects that are claimed to offset this may only end up lasting a few decades or less. For example, offsets created under the official Australian Carbon Credit Units (ACCU) scheme only have to prove they'll hold carbon for as little as twenty-five years. So offsets offer the wrong kind of storage, for too short a time, to deal with the carbon pollution that gets pumped out when we burn fossil fuels. This is why scientists are very clear that offsets can never be a substitute for deep and permanent reductions in carbon pollution.[25]

The second problem is technically called 'additionality', which basically describes whether or not there's a surplus of action when someone creates and sells an offset. For a carbon offset to be worth anything at all, it needs to account for pollution reduction or absorption *which would not have happened anyway*. Carbon-offset projects are supposed to prove that every tonne of benefit they deliver is entirely surplus, representing new emissions abatement that is only happening because of the project itself. This is a big issue for international carbon-offset projects, with a series of investigations in recent years exposing billions of dollars' worth of offsets as potentially junk.

Most prominently, in early 2023 an international coalition of journalists dropped a bomb on the global carbon market through an exposé questioning the quality of offsets certified by Verra. The Washington-based not-for-profit is the world's biggest agency for validating offsets purchased by major corporations in the voluntary market, including Australian brands like Qantas and Origin Energy.[26] The investigation claimed over 90 per cent of carbon offsets certified by Verra from rainforest-protection projects were essentially 'phantom' credits that delivered no additional carbon benefit. In some cases, rainforests which were supposed to be protected from logging were reported to have continued shrinking anyway. In others, credits were allegedly sold for protecting land that was never under threat to begin with.[27] Verra vigorously disputed the claims when they were first aired. But since then, its CEO has stood down and the company has started designing new rules for generating rainforest carbon offsets—which does rather suggest it spotted some problems in the end.

Lest we think this is just a case of problems with that particular firm, researchers from Europe, the United Kingdom and the United States recently collaborated on a large review of evaluations of over 2000 offset projects around the world. Their conclusion? The vast majority—88 per cent—of the total carbon abatement claimed by these projects 'does not constitute real emissions reductions'.[28] In most cases, this was because the promised activities either weren't delivered or weren't surplus to what would have happened without the offset project.

This isn't just an international issue, either. In 2022, the Australian Government commissioned an independent review of the ACCU scheme. While not identifying misconduct on the scale of the Verra scandal, the review made a wide-ranging series of recommendations to tighten up how offsets can be created in Australia. This included doing away with some methods, like paying farmers not to clear land they weren't going to clear anyway.

If dodgy carbon-offset projects don't create emissions abatement that is surplus to what would have happened anyway, they're doing worse than nothing. They're enabling companies to keep polluting as usual by providing a fig leaf to cover up their emissions. And they're doing this without even providing the modest benefit of temporarily storing that pollution in our land system like high-quality offsets would.

Finally, where are we going to source enough land to create all the offsets we'll need if we don't cut emissions, or the money to pay for these? Land to plant trees doesn't grow on … well, you get it. Land is a scarce commodity in its own right. A team of Australian researchers examined the climate pledges made by countries under the Paris Agreement, to look at how much these relied on carbon offsetting through land and forest restoration. They calculated that meeting these pledges with nature-based offsets would require over one billion hectares of land—more than the areas of the European Union (EU), India, South Africa and Turkey combined.[29]

Around half of the offsetting activity promised by these countries would require significant changes to how land is already being used; for example, planting trees on plots that are currently used for agriculture or industry. This raises huge question marks over whether those offsets will really be delivered, even if we set aside the science saying they cannot properly offset fossil fuel pollution. Drilling into individual countries' promises highlights just how much of a charade this is. The oil and gas behemoth Saudi Arabia has pledged to plant 10 billion trees within its borders over the coming decades, as part of restoring some 40 million hectares of degraded land. If they pull this off, it would represent a 4000 per cent increase in the amount of forested land across that famously arid country.[30] The term 'greenwashing' doesn't do justice to a promise like that; it's a straight-up fantasy.

And how much would all this cost? In Australia, extracting and burning coal, oil and gas in all their forms is projected to create just over 320 million tonnes of carbon pollution in 2030.[31] At the moment, the average price for an Australian Government–verified carbon offset is around $35. So offsetting all that pollution at this relatively modest price per tonne would cost about $11.2 billion—each and every year we continue emitting that much. If we scaled this up globally, offsetting all emissions from coal, oil and gas around the world could cost more than a trillion dollars a year.

It won't, of course. Because the companies and countries making net-zero promises that rely on masses of offsets don't really mean it. They're just saying the right things for now and hoping something else turns up—whether that's some entirely new technology for sucking up carbon pollution, or a 'Look over there!' crisis that takes our minds off all this greenie nonsense.

The bottom line is that burning fossil fuels produces far more carbon pollution than would ever occur naturally, so we can't offset all of this in nature. Getting back to a safe balance requires genuinely and permanently slashing pollution from coal, oil and gas close to zero. Only then will it be possible to find enough natural sinks to deal with the fraction left over from things we really can't avoid. Loading up on offsets to cancel out carbon pollution on paper is not the answer to tackling climate change, and it never can be. They're a last resort once we've done everything we can possibly do to drive it as close as possible to *real* zero.

The enthusiastic embrace of carbon accounting has had a direct impact on how much we expect governments, companies and—let's be honest—even ourselves to do about reducing harmful carbon pollution. Like a lot of other countries, Australia has signed up to the goal of reaching net zero by 2050. As a step towards that, the Albanese government has committed in law to cut carbon pollution by at least 43 per cent by the end of this decade. They don't like to emphasise it, but that's a 'net' target too. Carbon offsets are going to be doing a lot of heavy lifting to get us there, because the latest national emissions projections show total pollution from major sectors including transport, heavy industry and agriculture will barely budge between now and 2030.[32] The land sector is expected to grow as a carbon sink and soak up some of their excess so we can still hit our target—at least on paper. In fact, the *only* part of Australia's economy that is currently on track to make deep, genuine and permanent cuts to pollution by 2030 is the electricity sector, as we increasingly swap dirty coal-fired power for clean wind, solar and batteries.

That's not good enough. As UN Secretary-General António Guterres memorably put it, we need to be doing 'everything, everywhere, all at once' to cut carbon across every part of our economy and society.[33] Letting whole sectors off the hook for not taking real action as long as they buy offsets is a recipe for runaway global heating we won't be able to recover from.

One of the reasons offsets have become the go-to climate solution is that almost every major industry describes the pollution it produces as 'hard to abate'. This term has a technical meaning, describing emissions which are either impossible to reduce using the technologies we have available today, or are outrageously expensive to cut. Pollution from burping cows is hard to abate, because we don't yet have a viable solution that can affordably and effectively reduce the climate-warming contents of those impolite emissions. Some (but not all) of the emissions produced making cement are also hard to abate right now, because the new technologies needed to manufacture this important material more cleanly are still being developed and tested. High-quality offsets do have a limited, short-term role to play in helping these genuinely hard-to-abate sectors reduce their climate impact, while we work on real and lasting solutions.

But in the argy-bargy about emissions reduction, the practice of industries badging themselves as 'hard to abate' has acquired a political meaning which is totally removed from the technical one. This political connotation is roughly equivalent to saying, 'Nice government you got here. Be a shame if something happened to it.' You see, industries from manufacturing and mining to transport and construction claim they, too, are hard to abate. If mean old governments force them to actually cut emissions, they'll simply have no choice but to shut up shop, sack all their workers, cease contributing to our economic growth (that old chestnut) and leave the nation poor, cold and hungry. Governments have generally backed away from calling this industry bluff, settling instead for carbon 'abatement' through offsetting as a Potemkin-village solution big corporates can mostly live with. What a shame our planet can't.

The idea that most industries can't make real, significant cuts to their carbon pollution is, to be blunt, total bullshit. The Industry Energy Transitions Initiative delivered by the Melbourne-based Climateworks not-for-profit and a coalition of major Australian businesses spent three years examining decarbonisation pathways for some of our biggest-polluting domestic industries: iron and steel, aluminium, other metals, chemicals, and liquefied fossil gas. They concluded that, over time, it would be possible for these sectors to cut their carbon pollution by 92 per cent, effectively eliminating over 200 million tonnes of harmful $CO_2$ a year.[34] The study found that for most of the industries covered, this would require switching from gas to electricity, in combination with rolling out new production technologies to take advantage of zero-emission feedstocks like renewable hydrogen. The gas industry was the only one where achieving major cuts to pollution required shrinking production, which makes sense since a polluting fossil fuel will always be a polluting fossil fuel.

At the Climate Council, our own team has mapped out a pathway to see Australia cut emissions by 75 per cent across the entire economy by 2030, using technologies available now and changes we already know how to make. Our research shows that the big opportunities come from rolling out far more renewable electricity, so we can electrify most of our homes and businesses; cutting the use of gas and dramatically improving energy efficiency in industry and the built environment; using shared transport and walking and cycling for more trips, to cut down our reliance on private cars; and ending the clearing of land—particularly native forests—to stop releasing masses of stored carbon back into the atmosphere. These are things that can be done by big and small businesses, industries and communities across our cities and our regions, you and me and everyone we know.

Big polluters want us to think that it's all too hard, or we don't have the solutions. But deep, genuine and permanent cuts to carbon pollution are possible and Australia could be making far more of them—starting right now. Other wealthy and middle-income countries can, too. This is the only thing that can halt and then reverse dangerous global heating. By letting us believe otherwise, net-zero carbon accounting is doing more harm than good. That's why we need a new way of thinking about, and tracking, what counts as climate action.

## Genuine Cuts to Carbon Pollution: the Ones That Really Count

The headline change we should make is to set aside net carbon accounting for now and focus on making genuine, permanent cuts to pollution at every level—across individual businesses, sectors and nations. We are living in the critical decade for action, when there's still time to avoid the worst impacts of harmful climate change, but only if we slash carbon pollution. Net zero is for balancing the ledger when we're much closer to real zero emissions than we are today. So while it may make sense to return to it as a metric in the decades to come, for now, counting carbon pollution this way is simply preventing us from having a clear line of sight to how much progress we're really (not) making.

Focusing on genuine, permanent cuts to pollution is a straightforward change in theory and practice. If businesses can count their emissions for the purpose of offsetting them, they shouldn't have any problem working out whether they're going up or down without those carbon tokens. It would be a highly controversial change politically, because it would reveal that, like the emperor parading among his subjects with his sack out, many of our biggest companies don't have real plans for cutting most of their carbon pollution. So be it. Forcing them to be honest about this is an important part of getting on track for real reductions.

There is welcome movement in this direction overseas. In early 2024, the European Parliament agreed to new regulations completely banning the use of terms like 'carbon neutral' and 'climate positive' in advertising and packaging, where businesses rely on carbon offsetting as the basis for making these claims.[35] Anna Cavazzini, the chair of the EU parliamentary committee which led the push on these rules, was up-front in claiming this as a blow against the whole concept of carbon accounting. 'It should no longer appear that planting trees in the rainforest makes the industrial production of a car, the organisation of a soccer World Cup or the production of cosmetics climate neutral. This deception is now a thing of the past,' she told *The Guardian*.[36]

The EU's decision builds on a significant report released in late 2022 from an international expert panel brought together by the climate-crusading Guterres. The panel was tasked with advising how companies—and other entities, like regional and local governments—can make genuine climate commitments. These experts didn't sugar-coat things; their recommendations systematically dismantled much of what counts as corporate climate action today. Their final report declared companies cannot buy carbon offsets instead of focusing on cutting their emissions (sorry, ASX 200 types), nor can they claim to be working towards net zero while developing new coal, oil and gas projects (looking at you, Woodside and friends). They also called out corporations for putting off action until closer to 2050 at the expense of making real progress this decade, something almost every big firm making climate claims today is guilty of.[37]

To replace all this common greenwashing guff, the UN expert group's work set a clear and unambiguous standard for real climate action. That is: companies should have a plan to halve carbon pollution in real—not net—terms by 2030, and a longer-term pathway to permanently cut emissions by at least 90 per cent before 2050. High-quality carbon offsets may play a role in accounting for the remaining 10 per cent of pollution, but *only* for this small residual share. Importantly, too, the group said these targets should cover carbon pollution across a company's entire value chain—meaning those produced in their onsite operations, those linked to their use of electricity, *and* those created through the downstream use of their products. This is a big deal, because for some companies—particularly those producing coal, oil and gas—this end-use pollution can account for 70–90 per cent of their total carbon contribution.[38] Despite this, few are yet taking responsibility for doing anything about it.

Armed with this new test, there are lots of useful ways in which we can update Australian laws and policies to prioritise legitimate cuts to carbon pollution. First and foremost, we need strong anti-greenwashing laws which prevent companies from claiming they are taking real climate action unless their plans are consistent with at least halving their gross carbon pollution by 2030. This is not a standard every company will be willing or able to meet, but that's OK. The point here is to distinguish between those who are on the right pathway and those who aren't yet there, so that Aussie consumers and governments alike can tell them apart. This is essential for creating fair competition within sectors as firms start to decarbonise, because there is going to be some up-front cost involved in all this. Companies that do the right thing and invest now to transform their operations shouldn't be disadvantaged by competing firms free-riding on bogus climate claims.

In late 2023, the Australian Competition and Consumer Commission (ACCC) released new guidelines for business on making environmental and sustainability claims. Unfortunately, these are laughably weak because the ACCC can't make new laws, only enforce existing ones.[39] So the Australian Government needs to raise the bar for business by making new anti-greenwashing rules that align with international advice on what constitutes real climate action.

We also need to update the settings for national frameworks like the Safeguard Mechanism, which regulates carbon pollution from our 200-odd biggest emitters. In 2023, the Albanese government delivered important reforms that will require entities covered by this policy to start reducing their emissions year-on-year. That's a big step in the right direction, given the former Coalition government's version of the mechanism allowed pollution to keep rising.[40] But at the moment, the responsible corporations can achieve their annual cuts entirely through the use of carbon offsets, if they so choose. And it looks like many of them will—the government's own projections estimate that a little under two-thirds of the emissions reduction achieved by the scheme over the years to 2030 will come from offsets, not direct cuts to pollution.[41] That needs to change. Regulated entities should be required to cut their emissions each year in real terms, against tighter targets which put them on track to at least halve these by 2030.

Being realistic, this may require allowing a handful of industries currently covered by the policy to drop out of it and be given bespoke targets for cutting pollution in the near term. This would include industries like cement and aviation, where the pathway to significant and lasting emissions reduction isn't yet clear via the technologies we have on hand today. But while we're adjusting the coverage, the Safeguard Mechanism could be expanded to capture far more of the corporations which pump out carbon pollution across the Australian economy. At the moment, it only picks up individual facilities—like factories or coalmines—which produce more than 100 000 tonnes of carbon pollution a year. In parallel, the National Greenhouse and Energy Reporting (NGER) scheme requires corporations that produce more than 50 000 tonnes of pollution a year across their entire operations to track and report their emissions. Currently, they're not required by law to reduce these, only to tell us about them.

Together, these companies produced over 374 million tonnes of carbon pollution in 2022–23, or just over 80 per cent of Australia's emissions that year.[42] They're a true cross-section of our economy, spanning big retailers to banks, construction companies to tech firms, freight companies to universities and major healthcare providers.[43] Every one of these companies also has a role to play in cutting carbon pollution; indeed, many of them have already made their own voluntary net-zero commitments (too often based on offsets, but hey, at least they're on board with the mission). So a practical step that builds on Australia's existing policy frameworks would be to extend the Safeguard Mechanism's requirement for real, annual cuts to pollution to all these corporations as well. How could they possibly have a problem with that, when so many already claim to be well down the right path?

That's being facetious; they'd probably go absolutely spare. But many of these companies can do a lot to cut their pollution fairly easily because it comes from things like their use of energy, buildings and vehicles. These can be dealt with by switching to clean technologies, as Australia's largest supermarket chain is admirably demonstrating. Woolworths has set a target of cutting their operational and electricity emissions by 63 per cent by 2030 *without* the use of carbon offsets; in 2023 they reported being over half-way to this goal.[44] They're getting there by doing the practical things every company should be doing now: powering their operations with renewable electricity, improving energy efficiency across their stores, and swapping out petrol cars and trucks for zero-emission alternatives. This isn't just good for our climate, it's also good for business, as these changes help slash ongoing operating costs at a time of high and volatile energy and petrol prices.

Genuine voluntary action like that being taken by Woolies is very welcome, but too often it's the exception among big companies. Requiring real, ongoing cuts to pollution by every corporate entity covered by the NGER scheme would put them all on the same path, to see our national emissions drop much more steeply this decade.

Alongside these changes to hold corporations accountable for driving real cuts to carbon pollution, we need to apply the same standard to government targets. As I've explained, land-based offsets are playing a big part in our national plan to cut pollution by 43 per cent by 2030. If we excluded the land sector from the current projections, our emissions would be just 33 per cent lower—way off track.[45] We need to start accounting more transparently for our progress across the economy so we can face up to these kinds of gaps. That means setting science-aligned national targets for real emissions reduction, and accounting for any use of offsets separately to these. This is a change we can make ASAP: under the Paris Agreement, all countries are due to submit updated national targets for the coming decade by the end of 2025. Australia can set its upgraded target for 2035 in real terms first, with a clearly separate net number which captures our planned use of offsets. There are murmurings of interest internationally for all countries to move to this more up-front approach for the next set of national commitments. We could lead from the front for once in doing so, leaving behind our reputation as a climate laggard.

Taken together, these changes would completely overhaul how we define and account for climate action in Australia, and set a new expectation that *all* of our biggest companies will genuinely contribute to cutting carbon pollution. They're also mutually reinforcing: requiring more real action from corporations makes it possible for us to aim for, and achieve, deeper cuts as a nation. They'd be contentious, for sure—it would take a coalition of progressive support across our national parliament to put them in place. But they're also pretty hard for any company or political party that is serious about climate action to argue with. Who could be against telling Australians the truth about the real progress we're making on keeping them safe from runaway climate change? Or requiring big companies that are already promising to work towards net zero to prove it, and really deliver it? To oppose these changes would be to settle for obfuscation and greenwashing on a massive scale, siding with bad faith actors who want to keep poisoning our collective future while getting plaudits for being part of the solution. Anyone putting their hand up for that?

## FALSE ECONOMY

Is a billion dollars a lot of money? How about a trillion? The only sensible answer to that question is: it depends what you're spending it on. Doing anything of substance at a national scale often involves spending sums with that many zeros in them. It's an infuriating kink of public policy that people treat these big numbers as inherently good or bad, scary or insufficient, wasteful or worthy. They're not—it's all a question of what we get for the money, and what else we would or could be spending it on instead. A billion dollars is arguably great value if poured into boosting bulk billing or adding more teachers to classrooms. Perhaps less so if spent on self-congratulatory public ad campaigns, like the last Coalition government did.[46]

This relativity is particularly hard for people to hold in mind when considering the costs of tackling climate change. It's not your fault if you've ever been struck by the sums that get tossed around, and found yourself wondering whether we can afford it. There's a deep information asymmetry embedded in our current conversation that makes it hard even for well-informed people to weigh up the merits of the investments we need to make now.

The expected costs of climate action are very visible. We see them all the time in dense reports by international agencies, and creative modelling by consulting firms on the payroll of one peak body or another. There's no question that renewing our energy system, developing new clean industries, replacing our vehicles and retrofitting our homes is going to cost serious money. No honest advocate for climate action would suggest otherwise, and we shouldn't shy away from acknowledging this. But it's rare to get a clear and concise answer to the only question that helps us make sense of the sums that will be involved: *compared to what?* The costs we'll face doing business as usual, and the compounding costs of inaction that lead to escalating global heating, are nowhere near as visible as they need to be for us to make a proper comparison. So we get stuck fretting over how we can afford it, instead of recognising that accelerated action this decade is the best all-in bet we'll ever make. We need to get better at counting the costs on both sides of this ledger, so we can put to bed once and for all the dangerously bad idea that strong action now will cost us more than inaction.

To get us going, let's bite the bullet and put an expected price tag on transforming Australia into a zero-carbon economy. The most detailed analysis of this so far was done by a big collaboration of universities through the Net Zero Australia project, which delivered its findings in 2023. Nearly fifty researchers and modellers worked for over two years to nail down all the expected steps required to reach our long-term national target. Then they drilled into the details of delivering these. Usefully, the team looked at a range of possible futures spanning faster or slower uptake of renewable energy, when and how we replace our fossil fuel exports, and even whether we reach net zero before the middle of this decade. I won't leave you in suspense: they reckoned the cost of getting to net zero would range from $4.8 trillion to $5.1 trillion.[47] They estimated we'd spend that sum over the years to 2050. Since the science says Australia needs to reach net zero well before this, we should probably assume we'll need to spend something like this amount even sooner.

Those are big numbers that prompt quite a gulp the first time you see them, I know. But that's exactly the trap I'm talking about, because is that really a lot of money? Is it more, less or exactly the same as what we'd spend by chugging along our fossil-fuelled way with business as usual? It's only by costing the counterfactual that we can answer that question. Strap in, because there's lots of evidence BAU will have an eye-watering bill of its own.

What are the direct costs of continuing to power ourselves with fossil fuels across our homes, businesses, transport and industry? For starters, we'll need to build a bunch of new electricity generators, because the ones we've got are rapidly breaking down. There are eighteen coal-fired power generators around Australia. Only five were built this century; twelve of them are already more than thirty years old, when the safe operating life for plants like these is around forty-five years. It's a little-understood fact that we're not just racing to roll out more renewables now because of climate change. We're also on a serious deadline to get new power online before these great crumbling behemoths give up the ghost.

The CSIRO runs an annual review of how much it costs to build new energy infrastructure with different technologies. Their latest report, released at the end of 2023, was blunt in stating that wind and solar are hands down the cheapest way to do this now, and they're only set to get cheaper.[48] Delivering new, large-scale solar-generation capacity is nearly four times cheaper than new coal plants, and almost 30 per cent cheaper than new gas-generation capacity. So as we have to replace our current power generators over the next two decades, rebuilding with fossil fuels would mean paying a massive premium. And you can add to this the cost of building more coalmines and gas wells to source the fuel for these dirty generators as the supplies we've got run down. Woodside's proposed Scarborough gas project off the WA coast is slated to cost $16.5 billion all on its own, so that infrastructure doesn't come cheap either.

To pre-empt the internet nuke bros who say that mini-reactors are the answer to both our climate and our costing troubles here, the latest CSIRO report had an even blunter message on that. Based on real pricing from the only project actually being developed in the world, nuclear would be over five times more expensive to build than coal, and twenty times more expensive than large-scale solar. Oh, and it couldn't come online until sometime in the back half of the 2030s, after we've already been sitting in the dark for years because our coal plants crapped out without a replacement.

Alongside replacing all of our energy generators in the decades ahead, we'll also need to update or build from scratch almost everything we currently power with fossil fuels. The average age of a car in Australia is just under eleven years, and more than half of all households own two or more of them. That means on current trends, between now and 2050 most of us will buy at least four or five replacement cars (although ideally we wouldn't, since using shared transport and walking and cycling more often are great ways to quickly cut carbon pollution). The lifespan of our household appliances varies a lot, but it's also a decent bet you'll have to replace your home heating system at least once in the next couple of decades, and appliances like your stove and hot water heater several times over. Australians who own their homes move every eleven years on average, so if you're lucky enough to be in the market, you'll probably also buy at least one totally new property. At the moment, we build around 170 000 new homes a year, with an estimated 10 per cent of these being knockdown–rebuilds of existing properties. Over the next quarter-century, that means we'll add about 3.8 million more properties to our housing stock, and totally renew around another 425 000 of those standing today. All of this is money we're collectively going to spend one way or another, because we'll *need* these replacement cars, appliances and homes. And we don't have to find it all in government coffers; we'll finance much of it ourselves, just like we did for the stuff we own today. The only question is whether we spend this money in ways that keep cooking the planet, or on cleaner and safer alternatives.

The Net Zero Australia study was wise to this. Alongside the different decarbonisation options, they also looked at a business-as-usual scenario where Australia did not prioritise cutting emissions but just kept patching up our fossil fuel energy system with more of the same. They put the cost of this at $4.3 trillion out to 2050. You'll spot that this is not far off the $4.8 trillion bottom end of their price range for taking action to decarbonise. And that's the essential point. Our choice now is not to spend nothing or spend trillions moving beyond fossil fuels. Our choice is between pouring trillions down the drain clinging to coal, oil and gas and turbocharging dangerous heating in the meantime, or spending something like the same sum to set our economy and community up for our next era of clean prosperity. Which one seems like a better investment?

If you're such a fiscal hawk that you'd still favour a fossil-fuelled future because there may be *some* price gap, hold tight. These direct costs aren't the only ones we have to factor in. Climate change is already costing us a lot through its impacts on our weather, environment, livelihoods and lives, and these impacts are set to escalate dramatically if global heating continues unabated. Counting the future costs of today's inaction quickly tips that ledger deep into the red.

Extreme weather is the most direct and visible impact of climate change, and Australians will bear the brunt more than many. With a hotter climate comes an increased frequency and severity of heatwaves, turbocharged bushfires, ferocious storms, and huge swings between floods and droughts. The direct costs of all this extreme weather are particularly easy to count because the insurance industry keeps close track of them. Since the Black Summer bushfires ripped through Australia in 2019–20, the Insurance Council of Australia says its members have paid out more than $16.8 billion in disaster claims.[49] This bill includes $6 billion in claims just for the massive 2022 floods which left communities like Lismore and Ipswich underwater for more than a week—to date, the single largest insurance event in Australian history.

The frequency, severity and cost of unnatural disasters like these have tracked closely along with rising global temperatures in recent decades. In the United States, the National Centers for Environmental Information have been keeping a long-running tally on climate-related disasters that do more than $1 billion worth of damage. In the 1980s, there were thirty-three of these disasters—an average of 3.3 a year—with a total damage bill of US$213 billion. By the 2010s, this had risen to 131 disasters—an average of more than thirteen a year—costing just under US$1 trillion. The current decade looks set to easily top that record, as there were sixty-six disasters costing US$431 billion between 2020 and 2023 alone.[50] The Insurance Council of Australia's actuarial wonks project extreme weather will bring a local bill of $35 billion a year by 2050 if we keep on our current path.[51]

Some of Australia's biggest industries also face a much tougher future in a warming world. Last year, our agriculture sector produced $78 billion worth of goods like meat, vegetables, wheat and (crucially) wine.[52] This sector feeds us, along with millions of others overseas. Water is the universal resource all farmers rely on to keep animals thriving and fields producing. But at 1.5°C of warming, the frequency of major droughts globally is expected to double from once a decade to every five years; higher temperatures will make them even more common.[53] That could see huge tracts of today's productive land being abandoned because it's simply too unreliable to farm there. Tourism is another big earner and employer; about 2.2 million people a year visit the Great Barrier Reef alone. How many of those natural beauty seekers would sign up to scuba on a bleached and sludge-covered coral graveyard? With 2°C of warming, we could see a 99 per cent reduction in our remaining healthy reefs.[54] The Australian Conservation Foundation has estimated that roughly half the total value of Australia's economy comes from industries that are moderately to extremely reliant on a healthy environment.[55] Of course, we shouldn't be fussed about the GDP value of all this activity. But we should care a great deal about the jobs, essential goods and social benefits all those industries provide.

One more before we put down the accountant's pen. The extreme weather that climate change fuels is really bad for our physical and mental health, with bushfires and floods being particularly harmful. Remember the eerie smoke that invaded cities across eastern Australia for weeks during the Black Summer fires? This caused a big uptick in trips to GPs and emergency departments for heart and respiratory problems, along with hundreds more deaths than would otherwise have been expected during that time.[56] The smoke-related healthcare costs of those anxious few weeks have been estimated at over $1.9 billion.[57] With climate change fuelling more frequent megafires and longer fire seasons, this will keep ramping up those healthcare costs. So will the increased spread of infectious diseases like dengue, malaria and zika virus as a warmer climate increases the range for the mosquitoes which carry them; and the incidence of waterborne diseases every time our towns and cities flood.[58] Events like these also leave a lasting mark after the immediate crisis is over. Among Australians who have experienced an extreme weather disaster like a major fire or flood, over half report dealing with symptoms of depression in the aftermath, and more than 70 per cent say they've experienced symptoms of anxiety.[59]

There's literally no shortage of studies into the health risks and costs we face from harmful climate change. So much so, there has been a study into the availability of these studies. The 2023 *Lancet* Global Health Countdown report noted Australia is 'extremely well served by research on the health impacts of climate change', with researchers having produced about fifty studies per million people on the topic. This is more than double the number of studies available for any other region of the world.[60] If health research alone was the spur, we should be tackling climate change twice as fast as other countries.

I understand that putting it all down in one long tally like this can feel overwhelming. The price of inaction spins into the trillions in ways that can be hard to keep hold of. But it's honestly quite difficult to *over*estimate just how massive the costs and consequences of global heating will be if we don't get it under control soon. Australia at 2°C or 3°C of warming will be radically and catastrophically different from the one we know today.

We're currently not properly counting this cost anywhere when making decisions about investment in climate action. I talked earlier about the information asymmetry we all face when trying to decide if spending on clean energy, industry, buildings and transport stacks up. In my experience, this asymmetry gets worse, not better, the closer you are to actually making those decisions about what to spend, and where to spend it. Ministers and CEOs are directly responsible for balancing the books—whether that's the nation's budget or a corporate profit-and-loss. This can bring a narrow focus on managing those amounts which *have* to be counted in them. Costs that are uncertain, off on the horizon, or don't feel like they need urgent attention today, can easily slip down the priority list. This is especially true within Australia's weird political culture, which fetishises budget surpluses. (Did you know that none of the governments of the United States, France or Japan have been in the black for the past twenty years?) I've lost count of the number of times I've heard a politician say words to the effect of 'We'll cross that bridge when we come to it', to justify putting off budgeting for an expected but indefinite cost. Businesses do the same; you're probably even recognising the habit in your own household budget right now.

We need to make the future costs of inaction just as visible and visceral as the dollars we spend doing things today. Leaders need to be rewarded for making long-term investments which reduce the very real climate risks we face, not attacked for spending more on them. We all need to understand where our own future actions will fit into Australia's renewal, and how to make smart choices with these. Getting good, transparent numbers on the real cost of business as usual out there can help balance this debate, and to open the floodgates of necessary investment we could be making at every level this decade.

### Quantifying and Communicating the Costs of Business as Usual

Federal and state budgets are a yearly exercise in setting out government priorities and explaining plans to tackle big social challenges and community needs. Hardly anyone except a handful of nerds reads the budget papers cover-to-cover. But they play an important role in shaping 'the vibe' around what needs attention, and spotlighting emerging risks—like future changes in the value of Australia's exports, or our ageing population. One important way in which we can make the costs of climate inaction far more visible is for governments to estimate these in a decent degree of detail, and include the aggregated cost as a forward liability or risk in their annual accounts.

They could then show how the actions and investments they're taking now will reduce this expected liability by cutting climate risks—a useful way of neutralising any political accusations about profligate spending today.

Since its first federal budget in 2022, the Albanese government has begun including more specific reporting on climate investment, together with some high-level commentary on climate-related risks. This is a good start, but it doesn't give anything like a clear enough picture of the massive costs we'll face across our entire economy if we don't halt and then reverse global heating. Most state and territory governments don't yet include even this light-touch level of reporting in their own budgets, with Victoria being a laudable exception. Public accounting practice requires the preparation of a statement of risks alongside every government budget, but this usually only covers near-term risks that could change the financial picture over the forward estimates. Building out these risk statements with more detailed figures on the medium- and long-term costs of climate inaction would make much clearer what we stand to gain by going big on action and investment this decade.

We can also do better at counting the cost of carbon pollution when deciding if individual policies, projects and investments are a good idea. At the moment, the question of whether something will add to, or help cut, harmful carbon pollution rarely gets asked and costed in a systematic way by governments or businesses. When a department does an impact analysis, they consider direct costs such as the price of compliance for businesses, and direct benefits such as job creation. Depending on how much they want to juice the numbers, they'll sometimes also count a range of 'second-round effects', like changes in consumer spending or demand for government services. Corporate cost-benefit analyses take a similar approach. In these instances, the huge indirect costs of fuelling more harmful climate change and the real but diffuse benefits of reducing carbon pollution don't get a look in.

This is why, going forward, all government and corporate costings should be required to count the 'social cost of carbon'. This is a standard estimate of the price tag for each additional tonne of carbon pollution produced. On the flip side, it can also estimate the financial benefit of any action taken to cut it.

The replacement of our ageing electricity generators provides a useful example. The Eraring power station, located to the south of Newcastle, is Australia's largest coal-fired generator, producing about 12.7 million tonnes of pollution a year. It's set to close in 2025, and there's a race on to provide new capacity to cover its 3-gigawatt load. When weighing up the best technology to do this, a dollar value representing the social cost of carbon should be applied to every tonne of pollution a replacement coal-fired plant would pump out. That same dollar value would be applied in calculating the benefit of avoiding those emissions by building renewable generation capacity instead. This would mean that even if it were cheaper in the short term to replace Eraring with another coal-fired generator (the CSIRO says it's not), applying a social cost of carbon would help balance the ledger in favour of the cleaner option.

This approach has been standard in the United States and Canada for some time. Since coming to office, US President Joe Biden has both increased the rate used in federal assessments to US$51 per tonne of carbon pollution, and expanded the range of agencies required to use this in their costings. In 2021, the Australian Capital Territory became Australia's first jurisdiction to apply a social cost of carbon in its budget processes, but so far no-one else has followed their lead. Factoring in this cost should now be a standard requirement of all government policy and investment assessments, across our federal and state government agencies. Accounting requirements should be updated for private companies and the consultancies who sell them cost–benefit analysis services as well, to embed this in their decision-making.

As counting the social cost of carbon is made mandatory, this should be set at a rate which recognises how close we are to the point of no return with climate change. A 2022 study in the renowned research journal *Nature* pointed out that the latest science says catastrophic changes to the global climate could occur at much lower levels of global heating than previously thought. Because of this, the authors of the study argued the social cost of carbon should now be set at something like US$185, to reflect just how much any further pollution could cost us.[61]

Finally, Australians need a clear and simple explanation of how much of the required change could come from all of us simply making different choices when we're buying, building and upgrading the stuff we use everyday in the years ahead. Doing a home reno? Put in some insulation or double glazing while you're at it, to improve your home's energy efficiency. Buying a new hot-water service? Make it an efficient electric one to cut your gas bill and emissions.

Considering replacing your family's second runabout car? Look at an electric vehicle, or try using shared transport, walking or cycling for a couple of weeks to see if you can do without it, because all of these options will save on petrol and pollution. Helping Australians to see and understand how our own actions can either be a continuing part of the problem or an important piece of the solution, will make the costs and steps involved in decarbonising our economy feel much less scary, and much more achievable.

Climate groups often worry about highlighting individual action for fear of taking the focus off the massive fossil fuel corporations who are most to blame, or getting people offside. But if we're all chipping into that $4.3 trillion bill Net Zero Australia estimated for BAU anyway, isn't it a positive message to know that our own choices and actions can help redirect this investment towards the cleaner options that will make a difference now? And doesn't it take some of the political heat out of the conversation if everyone understands that a big share of the trillions that need to be invested is money we—governments, businesses and households alike—would have to spend one way or the other? Governments can lead the conversation by running large-scale public information campaigns which explain to the community the role we can all play, and civil society organisations should get much more active in this space, too. We need to contest the public space that has been far too dominated by opponents of action throwing around big numbers and scaremongering about our ability to afford it. The fossil fuel industry and its political cheerleaders *want* us to think the price of climate action is too high, because every day we delay, our BAU spending keeps flowing directly to their bottom lines. Helping Australians understand that our everyday choices can be part of the change will take the bite out of this big scare.

Humans are notorious for our bias towards short-term thinking. There's really no better example of it than our obsession with balancing budgets now at the expense of investing for a safe climate future, ignoring the real costs that come with continuing down the path of unchecked pollution. Evening out the information asymmetry in our current public debate about climate action is an important way of breaking through this bias, so we can recognise that it is *inaction* that carries the heaviest bill now.

~

Some truths are hard. That doesn't absolve us of the obligation to face them.

One hard truth about tackling climate change is that we can't do it half-arsed from here. We've left it too long, let the risks mount too high, for gradual and incremental change to be an option anymore.

That's why the three climate clangers explored in this book are such a problem, They're all handbrakes on momentum at a time when we need to throw the switch to warp speed. Transition gradually, lest we disrupt the forward march of GDP growth; cut lightly, because offsets will make the problem go away; spend carefully, that we might never be accused of acting beyond our means. Each is an insidiously subtle reason for delay, all the more dangerous because those with the power to drive the change we need now don't recognise them as such. Enough. We're out of time to let bad ideas restrain our actions a moment longer. We need to consciously put them aside.

Another hard truth is that we can't fix this problem by relying on the same economic, political and financial structures that got us into this climate mess in the first place. We're going to have to do things that straight up disrupt them. Like putting power onto the roofs and into the hands of homeowners, instead of into the profit forecasts of massive multinational fossil fuel corporations. Like forcing businesses to prioritise cutting pollution over providing their shareholders with more dirty dividends. Like investing now and being OK with the fact that the best returns we'll get come from the harms we *don't* end up seeing. We can't and won't do these necessary things if we keep stubbornly clinging onto old ideas about how the world should work. We need the creativity to imagine a different Australia, and the courage to go out and build it.

The hardest truth of all is that we face the responsibility and the test of living through the last years where it may actually be possible to fix this mess. Our actions this decade will make the difference between halting harmful warming or seeing it accelerate out of all control. We can be angry at those who ignored the imperative to act earlier, and awed at the size of the task that now falls to us. But our only real choice is to roll up our sleeves and fix it: by rapidly phasing out fossil fuels, renewing our energy system, and transforming how we make things, build things and move around. These climate clangers can't be our excuses any longer, the intellectual crutches we are using to justify more delay.

The way we understand a problem defines the solutions we can imagine to solve it. It's time to think differently about what a thriving zero-carbon economy looks like, the bar we set for real climate action and the dividends of strong action now. Changing how we think about tackling climate change can make possible things that feel unimaginable right now. It can unlock the political and social support that underpins all lasting change, and accelerate Australia's decarbonisation during this genuinely make-or-break decade. Once we put aside the bad ideas that are blocking real action today, together we can get on with doing what's necessary.

# ACKNOWLEDGEMENTS

Thank you to the diverse array of deeply committed Australian politics and climate movement friends and colleagues who have informed the ideas in this book through interesting conversations and lively debates in the past few years. Particular thanks to Professor Lesley Hughes, Amanda McKenzie, Dr Martin Rice, Dinah Arndt and Dr Adam Triggs for their challenging comments and helpful feedback throughout the drafting process.

I am very grateful to In the National Interest series editor Greg Bain for the opportunity to contribute to this conversation alongside such an illustrious crew of fellow authors, and to Paul Smitz for his thoughtful and considered approach to bringing out the best in the manuscript.

A special thank you to *mes proches* for your patience in creating the space for me to bring my day job into our weekends and after-hours for a little while. And finally, thank you to the National Library of Australia and all the wonderful humans who work within it, for providing a peaceful oasis where ideas can take root and flourish.

# NOTES

1 United Nations Environment Program, *Emissions Gap Report 2023: Broken Record—Temperatures Hit New Highs, Yet World Fails to Cut Emissions (Again)*, 20 November 2023, https://www.unep.org/resources/emissions-gap-report-2023 (viewed March 2024).

2 R Newman and I Noy, 'The Global Costs of Extreme Weather That Are Attributable to Climate Change', *Nature Communications*, September 2023, https://www.nature.com/articles/s41467-023-41888-1 (viewed March 2024).

3 S&P Global, 'Australia: Mining By the Numbers, 2021', 8 February 2022, https://www.spglobal.com/marketintelligence/en/news-insights/research/australia-mining-by-the-numbers-2021 (viewed March 2024).

4 Climate Council, 'Australia's Clean Industry Future: Making Things Here in a Net Zero World', 2023, https://www.climatecouncil.org.au/wp-content/uploads/2023/03/CC_MVSA0350-CC-Report-Industrial-Decarbonisation_V8-FA-Screen-Single.pdf (viewed March 2024).

5 Chevron Australia Holdings, 'Economic Contribution Report for Year Ended 31 December 2021', 2022, https://australia.chevron.com/-/media/australia/publications/documents/tax-transparency-factsheet.pdf (viewed March 2024).

6 Chevron Corporation, 'Chevron Announces Fourth Quarter 2021 Results', 28 January 2022, https://www.chevron.com/-/media/chevron/stories/documents/4Q21-earnings-press-release.pdf (viewed March 2024).
7 R Bousso and S Nasralla, 'Shell 2022 Profit More than Doubles to Record $40 Bln', *Reuters*, 2 February 2023, https://www.reuters.com/business/energy/shell-makes-record-40-billion-annual-profit-2023-02-02 (viewed March 2024).
8 Department of the Treasury, *Intergenerational Report 2023: Australia's Future to 2063*, 2023, https://treasury.gov.au/sites/default/files/2023-08/p2023-435150.pdf (viewed March 2024).
9 J Chalmers, 'Critical Minerals: A Chance to Secure the Future', *The Australian*, 25 November 2022, https://ministers.treasury.gov.au/ministers/jim-chalmers-2022/articles/opinion-piece-critical-minerals-chance-secure-future (viewed March 2024).
10 J Vogel and J Hickel, 'Is Green Growth Happening? An Empirical Analysis of Achieved Versus Paris-Compliant CO2–GDP Decoupling in High-Income Countries', *The Lancet Planetary Health*, September 2023, https://www.thelancet.com/journals/lanplh/article/PIIS2542-5196(23)00174-2/fulltext (viewed March 2024).
11 Ibid.
12 K Hubacek, X Chen, K Feng, T Wiedmann and Y Shan, 'Evidence of Decoupling Consumption-Based $CO_2$ Emissions from Economic Growth', *Advances in Applied Energy*, vol. 4, November 2021, https://www.sciencedirect.com/science/article/pii/S2666792421000664 (viewed March 2024).
13 Zidong An, Nathalie Gonzalez Prieto, Prakash Loungani and Saurabh Mishra, 'Does Growth Create Jobs? Evidence

for Advanced and Developing Economies', *IMF Research Bulletin*, September 2016, https://www.elibrary.imf.org/view/journals/026/2016/003/article-A002-en.xml (viewed March 2024).

14 All employment figures quoted in this section sourced from: Australian Bureau of Statistics, 'Labour Force, Australia, Detailed—Table 06: Employed Persons By Industry Sub-Division of Main Job (ANZSIC) and Sex', 2023, https://www.abs.gov.au/statistics/labour/employment-and-unemployment/labour-force-australia-detailed (viewed March 2024). Author's calculations.

15 Net Zero Australia, 'Final Modelling Results', 19 April 2023, https://www.netzeroaustralia.net.au/final-modelling-results (viewed March 2024).

16 Jobs and Skills Australia, *The Clean Energy Generation: Workforce Needs for a Net Zero Economy*, 2023, https://www.jobsandskills.gov.au/studies/clean-energy-capacity-study (viewed March 2024).

17 Australian Bureau of Statistics, 'Labour Force, Australia, November 2023—Table 1: Labour Force Status by Sex, Australia—Trend, Seasonally Adjusted and Original, November 2023, https://www.abs.gov.au/statistics/labour/employment-and-unemployment/labour-force-australia (viewed March 2024).

18 Author's calculations based on data from Australian Bureau of Statistics, 2023.

19 Australian Bureau of Statistics, 'Multiple Job Holders—September 2023', 2023, https://www.abs.gov.au/statistics/labour/jobs/multiple-job-holders/latest-release#multiple-job-holding-over-time (viewed March 2024).

20 See, for example: EM Curtis, DG Garrett, EC Ohrn, KA Roberts and JC Suárez Serrato, 'Capital Investment and Labour Demand', National Bureau of Economic

Research, June 2022, https://www.nber.org/papers/w29485 (viewed March 2024); and P Aghion, C Antonin, S Bunel and X Jaravel, 'Modern Manufacturing Capital, Labour Demand and Product Market Dynamics: Evidence from France', November 2022, https://sciencespo.hal.science/hal-03943312/document (viewed March 2024).

21 Woodside Energy, *Submission to Review of the Petroleum Resource Rent Tax*, February 2017, https://treasury.gov.au/sites/default/files/2019-03/R2016-001_Woodside.pdf (viewed March 2024); and Woodside Energy, *2022 Annual Report*, 2022, https://www.woodside.com/docs/default-source/investor-documents/major-reports-(static-pdfs)/2022-annual-report/annual-report-2022.pdf (viewed March 2024).

22 Ibid.

23 Elizabeth Morison, 'Climate of the Nation 2023: Tracking Australia's Attitudes Towards Climate Change and Energy', The Australia Institute, 13 September 2023, https://australiainstitute.org.au/report/climate-of-the-nation-2023 (viewed March 2024).

24 World Bank, 'Record High Revenues from Global Carbon Pricing Near $100 Billion', press release, 23 May 2023, https://www.worldbank.org/en/news/press-release/2023/05/23/record-high-revenues-from-global-carbon-pricing-near-100-billion (viewed March 2024).

25 S Fankhauser, SM Smith, M Allen et al., 'The Meaning of Net Zero and How to Get It Right', *Nature Climate Change*, vol. 12, January 2022, https://www.nature.com/articles/s41558-021-01245-w (viewed March 2024).

26 G Readfern, 'Qantas, Origin and Other Australian Companies Urged to Check Effectiveness of Overseas Rainforest Carbon Credits', *The Guardian*, 20 January

2023, https://www.theguardian.com/australia-news/2023/jan/20/qantas-origin-and-other-australian-companies-urged-to-check-effectiveness-of-overseas-rainforest-carbon-credits (viewed March 2024).

27 P Greenfield, 'More than 90% of Rainforest Carbon Offsets By Biggest Certifier Are Worthless, Analysis Shows', *The Guardian*, 19 January 2023, https://www.theguardian.com/environment/2023/jan/18/revealed-forest-carbon-offsets-biggest-provider-worthless-verra-aoe (viewed March 2024).

28 B Probst, M Toetzke, A Kontoleon, L Diaz Anadon and VH Hoffmann, 'Systematic Review of the Actual Emissions Reductions of Carbon Offset Projects across All Major Sectors', working paper, ETH Zurich, 2023, https://www.research-collection.ethz.ch/handle/20.500.11850/620307 (viewed March 2024).

29 A Self, R Burdon, J Lewis, P Riggs and K Dooley, *Land Gap Report Briefing Note: 2023 Update*, 2023, https://landgap.org/2023/update (viewed March 2024).

30 Ibid.

31 Department of Climate Change, Energy, the Environment and Water, 'Australia's Emissions Projections 2023', 2023, https://www.dcceew.gov.au/climate-change/publications/australias-emissions-projections-2023 (viewed March 2024).

32 Ibid.

33 A Guterres, speech at the launch of the *Synthesis Report* of the Intergovernmental Panel on Climate Change, Interlaken, 20 March 2023, https://www.un.org/sg/en/content/sg/statement/2023-03-20/secretary-generals-video-message-for-press-conference-launch-the-synthesis-report-of-the-intergovernmental-panel-climate-change (viewed March 2024).

34 Australian Industry Energy Transitions Initiative, 'Pathways to Industrial Decarbonisation: Positioning Australian Industry to Prosper in a Net Zero Global Economy', February 2023, https://www.cefc.com.au/insights/market-reports/australian-industry-energy-transitions-initiative (viewed March 2024).

35 European Parliament, 'Final Vote on Banning Greenwashing and Misleading Product Information', 15–18 January 2024, https://www.europarl.europa.eu/news/en/agenda/briefing/2024-01-15/2/final-vote-on-banning-greenwashing-and-misleading-product-information (viewed March 2024).

36 P Greenfield, 'EU Bans "Misleading" Environmental Claims That Rely on Offsetting', *The Guardian*, 18 January 2024, https://www.theguardian.com/environment/2024/jan/17/eu-bans-misleading-environmental-claims-that-rely-on-offsetting#:~:text=Terms%20such%20as%20%E2%80%9Cclimate%20neutral,crackdown%20on%20misleading%20environmental%20claims (viewed March 2024).

37 United Nations' High-Level Expert Group on the Net Zero Emissions Commitments of Non-State Entities, *Integrity Matters: Net Zero Commitments by Businesses, Financial Institutions, Cities and Regions*, 2022, https://www.un.org/sites/un2.un.org/files/high-level_expert_group_n7b.pdf (viewed March 2024).

38 Dr Paul Griffin, *The Carbon Majors Database: CDP Carbon Majors Report 2017*, Carbon Disclosure Project, July 2017, https://cdn.cdp.net/cdp-production/cms/reports/documents/000/002/327/original/Carbon-Majors-Report-2017.pdf (viewed March 2024).

39 Australian Competition and Consumer Commission, *Making Environmental Claims: A Guide for Business*,

12 December 2023, https://www.accc.gov.au/about-us/publications/making-environmental-claims-a-guide-for-business (viewed March 2024).

40 A Morton, 'A 60% Rise in Industrial Emissions Points to Failure of Coalition's "Safeguard Mechanism"', *The Guardian*, 12 February 2020, https://www.theguardian.com/australia-news/2020/feb/12/a-60-rise-in-industrial-emissions-points-to-failure-of-coalitions-safeguard-mechanism#:~:text=RepuTex%20found%20emissions%20in%20sectors,government%20is%20acting%20on%20emissions (viewed March 2024).

41 Department of Climate Change, Energy, the Environment and Water, 'Australia's Emissions Projections 2023', 2023, https://www.dcceew.gov.au/climate-change/publications/australias-emissions-projections-2023 (viewed March 2024).

42 Author's calculations based on: Clean Energy Regulator, 'Corporate Emissions and Energy Data 2022–23', 28 February 2024, https://www.cleanenergyregulator.gov.au/NGER/National%20greenhouse%20and%20energy%20reporting%20data/Corporate%20emissions%20and%20energy%20data/corporate-emissions-and-energy-data-2022-23 (viewed March 2024); and Department of Climate Change, Energy, the Environment and Water, 'National Greenhouse Gas Inventory Quarterly Update', June 2023, https://www.dcceew.gov.au/climate-change/publications/national-greenhouse-gas-inventory-quarterly-update-june-2023#:~:text=Emissions%20for%20the%20year%20to,compared%20with%20the%20previous%20year (viewed March 2024).

43 Note that there is overlap in coverage between the Safeguard Mechanism–regulated entities and corporate groups required to report under the National Greenhouse

and Energy Reporting (NGER) scheme. In 2021–22, 417 corporate groups met the threshold for reporting under the NGER scheme. This figure included the owners of all 219 facilities covered by the Safeguard Mechanism.

44 Woolworths Group, *Annual Report 2023*, 2023, https://www.woolworthsgroup.com.au/au/en/investors/our-performance/reports.html (viewed March 2024).

45 Author's calculations based on: Department of Climate Change, Energy, the Environment and Water, 'National Greenhouse Gas Inventory Quarterly Update', June 2023, https://www.dcceew.gov.au/climate-change/publications/national-greenhouse-gas-inventory-quarterly-update-june-2023#:~:text=Emissions%20for%20the%20year%20to,compared%20with%20the%20previous%20year (viewed March 2024).

46 A Albanese, 'Scott Morrison Adds Billion Dollar Advertising Bill to Trillion Dollar Debt', Parliament of Australia, 8 January 2021, https://parlinfo.aph.gov.au/parlInfo/search/display/display.w3p;query=Id%3A%22media%2Fpressrel%2F7746553%22;src1=sm1 (viewed March 2024).

47 Net Zero Australia, 'Final Modelling Results', 19 April 2023, https://www.netzeroaustralia.net.au/final-modelling-results (viewed March 2024).

48 CSIRO, *GenCost 2023–24: Consultation Draft*, 2023, https://www.csiro.au/en/research/technology-space/energy/energy-data-modelling/gencost (viewed March 2024).

49 Insurance Council of Australia, *Climate Change Roadmap: Towards a Net-Zero and Resilient Future*, 2023 Update, 2023, https://insurancecouncil.com.au/issues-in-focus/climate-change-action (viewed March 2024).

50 National Centers for Environmental Information, 'Billion-Dollar Weather and Climate Disasters', February 2024, https://www.ncei.noaa.gov/access/billions/summary-stats#temporal-comparison-stats (viewed March 2024).

51 Insurance Council of Australia, *Insurance Catastrophe Resilience Report 2021–22*, 2022, https://insurancecouncil.com.au/wp-content/uploads/2022/09/20683_ICA_Final_WebOptimised.pdf (viewed March 2024).

52 Department of Agriculture, Fisheries and Forestry—Australian Bureau of Agricultural and Resource Economics, 'Agricultural Overview', 2023, https://www.agriculture.gov.au/abares/research-topics/agricultural-outlook/agriculture-overview (viewed March 2024).

53 A Wait and K Meagher, 'Climate Change Means Australia May Have to Abandon Much of Its Farming', *The Conversation*, 6 September 2021, https://theconversation.com/climate-change-means-australia-may-have-to-abandon-much-of-its-farming-166098 (viewed March 2024).

54 Intergovernmental Panel on Climate Change, *Global Warming of 1.5°C*, 2018, https://www.ipcc.ch/sr15/ (viewed March 2024).

55 Australian Conservation Foundation, *The Nature-Based Economy: How Australia's Prosperity Depends on Nature*, 5 September 2022, https://www.acf.org.au/nature-based-economy-report (viewed March 2024).

56 N Borchers Arriagada, AJ Palmer, D Bowman, GG Morgan, BB Jalaludin and FH Johnston, 'Unprecedented Smoke-Related Health Burden Associated with the 2019–20 Bushfires in Eastern Australia', *The Medical Journal of Australia*, vol. 213, no. 6, 23 March 2020,

https://www.mja.com.au/journal/2020/213/6/unprecedented-smoke-related-health-burden-associated-2019-20-bushfires-eastern (viewed March 2024).

57 Z Ademi, E Zomer, C Marquina, P Lee, S Talic, Y Guo and D Liew, 'The Hospitalisations for Cardiovascular and Respiratory Conditions, Emergency Department Presentations and Economic Burden of Bushfires in Australia between 2021 and 2030: A Modelling Study', *Current Problems in Cardiology*, vol. 48, no. 1, January 2023, https://www.sciencedirect.com/science/article/abs/pii/S0146280622003139?via%3Dihub (viewed March 2024).

58 JC Semenza, J Rocklöv and KL Ebi, 'Climate Change and Cascading Risks from Infectious Disease', *Infectious Diseases and Therapy*, vol. 11, no. 4, August 2022, https://www.ncbi.nlm.nih.gov/pmc/articles/PMC9334478 (viewed March 2024).

59 Climate Council, *Climate Trauma: The Growing Toll of Climate Change on the Mental Health of Australians*, 2023, https://www.climatecouncil.org.au/resources/climate-trauma (viewed March 2024).

60 PJ Beggs and Y Zhang, '*The Lancet* Countdown on Health and Climate Change: Australia a World Leader in Neglecting Its Responsibilities', *The Medical Journal of Australia*, 20 November 2023, https://www.mja.com.au/journal/2023/219/11/lancet-countdown-health-and-climate-change-australia-world-leader-neglecting (viewed March 2024).

61 K Rennert, F Errickson, BC Prest et al., 'Comprehensive Evidence Implies a Higher Social Cost of $CO_2$', *Nature*, vol. 610, 2022, https://www.nature.com/articles/s41586-022-05224-9 (viewed March 2024).

## IN THE NATIONAL INTEREST

**Other books on the issues that matter:**

David Anderson *Now More than Ever: Australia's ABC*
Bill Bowtell *Unmasked: The Politics of Pandemics*
Michael Bradley *System Failure: The Silencing of Rape Survivors*
Melissa Castan & Lynette Russell *Time to Listen: An Indigenous Voice to Parliament*
Inala Cooper *Marrul: Aboriginal Identity & the Fight for Rights*
Kim Cornish *The Post-Pandemic Child*
Samantha Crompvoets *Blood Lust, Trust & Blame*
Satyajit Das *Fortune's Fool: Australia's Choices*
Richard Denniss *Big: The Role of the State in the Modern Economy*
Rachel Doyle *Power & Consent*
Jo Dyer *Burning Down the House: Reconstructing Modern Politics*
Wayne Errington & Peter van Onselen *Who Dares Loses: Pariah Policies*
Gareth Evans *Good International Citizenship: The Case for Decency*
Paul Farrell *Gladys: A Leader's Undoing*
Kate Fitz-Gibbon *Our National Shame: Violence against Women*
Paul Fletcher *Governing in the Internet Age*
Carrillo Gantner *Dismal Diplomacy, Disposable Sovereignty: Our Problem with China & America*
Jill Hennessy *Respect*
Lucinda Holdforth *21st-Century Virtues: How They Are Failing Our Democracy*
Simon Holmes à Court *The Big Teal*
Andrew Jaspan & Lachlan Guselli *The Consultancy Conundrum: The Hollowing Out of the Public Sector*

*(continued from previous page)*

Andrew Leigh *Fair Game: Lessons from Sport for a Fairer Society & a Stronger Economy*
Ian Lowe *Australia on the Brink: Avoiding Environmental Ruin*
John Lyons *Dateline Jerusalem: Journalism's Toughest Assignment*
Richard Marles *Tides that Bind: Australia in the Pacific*
Fiona McLeod *Easy Lies & Influence*
Michael Mintrom *Advancing Human Rights*
Louise Newman *Rape Culture*
Martin Parkinson *A Decade of Drift*
Isabelle Reinecke *Courting Power: Law, Democracy & the Public Interest in Australia*
Abul Rizvi *Population Shock*
Kevin Rudd *The Case for Courage*
Don Russell *Leadership*
Scott Ryan *Challenging Politics*
Ronli Sifris *Towards Reproductive Justice*
Kate Thwaites & Jenny Macklin *Enough Is Enough*
Simon Wilkie *The Digital Revolution: A Survival Guide*
Carla Wilshire *Time to Reboot: Feminism in the Algorithm Age*
Campbell Wilson *Living with AI*